Javad

Naz

Khalid

Evaluation of Extraction... ducts

Sumera Javad
Kanwal Naz
Sana Khalid

Evaluation of Extraction Method for Household Herbal Products

LAP LAMBERT Academic Publishing

Impressum / Imprint

Bibliografische Information der Deutschen Nationalbibliothek: Die Deutsche Nationalbibliothek verzeichnet diese Publikation in der Deutschen Nationalbibliografie; detaillierte bibliografische Daten sind im Internet über http://dnb.d-nb.de abrufbar.
Alle in diesem Buch genannten Marken und Produktnamen unterliegen warenzeichen-, marken- oder patentrechtlichem Schutz bzw. sind Warenzeichen oder eingetragene Warenzeichen der jeweiligen Inhaber. Die Wiedergabe von Marken, Produktnamen, Gebrauchsnamen, Handelsnamen, Warenbezeichnungen u.s.w. in diesem Werk berechtigt auch ohne besondere Kennzeichnung nicht zu der Annahme, dass solche Namen im Sinne der Warenzeichen- und Markenschutzgesetzgebung als frei zu betrachten wären und daher von jedermann benutzt werden dürften.

Bibliographic information published by the Deutsche Nationalbibliothek: The Deutsche Nationalbibliothek lists this publication in the Deutsche Nationalbibliografie; detailed bibliographic data are available in the Internet at http://dnb.d-nb.de.
Any brand names and product names mentioned in this book are subject to trademark, brand or patent protection and are trademarks or registered trademarks of their respective holders. The use of brand names, product names, common names, trade names, product descriptions etc. even without a particular marking in this work is in no way to be construed to mean that such names may be regarded as unrestricted in respect of trademark and brand protection legislation and could thus be used by anyone.

Coverbild / Cover image: www.ingimage.com

Verlag / Publisher:
LAP LAMBERT Academic Publishing
ist ein Imprint der / is a trademark of
OmniScriptum GmbH & Co. KG
Heinrich-Böcking-Str. 6-8, 66121 Saarbrücken, Deutschland / Germany
Email: info@lap-publishing.com

Herstellung: siehe letzte Seite /
Printed at: see last page
ISBN: 978-3-659-75031-1

Zugl. / Approved by: Lahore, Lahore College for Women University, Dissertation, 2014

Evaluation of Extraction Method for Household Herbal Products

LIST OF CONTENT

Contents	Page No.
Chapter no. 01	
Introduction	02
Chapter no. 02	
Review of literature	06
Chapter no. 03	
Materials and methods	17
Chapter no. 04	
Results	21
Discussion	48
References	51

INTRODUCTION

Traditional herbal medicines are defined as a natural source of treating illness which are basically derived from untreated plant based substances. These herbal medicines have a very little industrial processing and even nil processing in some cases. These medicines are usually used locally with some household practices (Kaptchuk & Tilburt, 2008).

Recently the WHO has defined traditional or herbal drugs) as the plant extracts or modifications of plant extracts which were in use since long time in the human history and development, and still persist for use today.

Traditional medicine is the synthesis of medical treatment of disease whose practice is done by the physicians from their local system of medicine. Traditional herbal drugs or preparations contain extracts of plants, a little bit of minerals as well as some organic matter etc. The use of plant extracts as medicines has been recorded in Syrian, Chinese, Greek and Indian history from old times even 5000years back. The classical Indian texts consist of Atharvaveda, Rig-Veda, Samhita and Charka. Basically herbal medicines are revival of healing agents from very far off corners of history of plant based healing agents. It is considered as Scientific heritage for plant scientists as well as for local practitioners (Kamboj, 2000).

75-80 % of the population of developing countries still depend on the herbal medicine and derived products (Kamboj, 2000). People generally have belief that there are lesser side effects of the herbal medicines. People also rely on such medicines because they are cheap as well as readily available (Gupta & Raina, 1998). The World Health Organization (WHO) has also claimed that usage of herbal medicine in world population is 2-3 times more prevalent as compared to conventional drugs (Evans, 1994).

For centuries, Tea and coffee have been the most admired and trendy beverages, basically due to their pleasing taste and refreshment effects (Alan & Iris, 2004). Tea and coffee are in common use at work places because they induce wakefulness and overcome the mental and physical exertion. These both beverages are usually used as hot infusion. It has been reported in a number of research papers that tea and coffee contain biologically active secondary plant metabolites i.e., polyphenols, flavonoids, alkaloids and tannins etc. these plant secondary metabolites are considered to be very important for human health because they have anti aging, anticancer and anti microbial effects. It has also been reported that regular use of tea or coffee can decrease threat of development of many chronic diseases. A few scientists also reported some of the side effects of the use of tea and coffee however, these beverages are accepted as beneficial overall (Al- Rasbi & Khan, 2013). Joshanda, a herbal product which is taken as tea and is considered as a remedy for some disorders. Joshanda is also analyzed for its acetyl salicylic acid and antioxidant characteristic. Due to the presence of these chemicals, it is proved as a beneficial tonic for general health (Soomro et al., 2011).

Scientific methods turned out to be more superior and preferred techniques in early 19th century, after this revolution people started to call the practice of botanical healing as quackery. But with the increasing knowledge of side effects of conventional medicines as well as the increased resistance in the pathogens against these diseases, once again trend of people began to transfer from conventional medicines towards the herbal products as natural health. The important step towards the recognition of use of herbal medicines was the establishment of the office of Alternative Medicine by the National Institute of Health USA, in 1992. The use of herbal drugs was further enhanced and strengthened worldwide when the WHO encouraged developing countries to use traditional herbal medicines (Winslow & Kroll, 1998).

Extraction is the very basic step in formulation of herbal medicines and it involves the partition of medicinally active components of plant from the inactive, inert or undesired portion by using the solvents of different polarity. The products so acquired from plants are impure liquids, semisolids or even powdered extracts for oral or external use. These primary products may be infusions, decoctions, tinctures, pilular or fluid extracts. These herbal preparations are known as galenicals, due to Galen, the second century Greek physician. The extract attained in this way may be prepared for use as a medicinal agent directly in the form of fluid extracts and tinctures and the most common example is tea, coffee and traditional Qehwa etc. in the next steps these primary extracts may be further processed to be included in any dosage form i.e., capsules or tablets. Even in modern botany, phytochemistry and pharmacy, there are techniques to fractionate and isolate individual chemical units such as stevioside, hyoscine, ajmalicine, and vincristine etc. So it becomes clear that standardization of extraction procedures is very significant for the final quantity as well as quality of the herbal drug and it is also true even for household herbal products (Handa *et al.*, 2008).

History of microwaves started from its use in World War II with the development of radar technology, which led to its domestic application in microwave ovens. While its use for research purposes started in late 70s as a heating source for acid digestions (Abu Samara *et al.*, 1975). Microwaves in fact are electromagnetic radiations whose frequency ranges from 0.3 to 300 GHz (Camel, 2000). Due to their electromagnetic nature, microwaves have electric as well as magnetic fields which are perpendicular to each other. There are two reasons of heating due to electric field i.e., dipolar rotation and ionic conduction at the same time (Thue'ry, 1992; Demesmay & Olle, 1993; Sinquin *et al.*, 1993). As far as dipolar rotation is concerned, it is due to the alignment of dipolar molecules on the electric field in the solvent and the solid sample. This orientation and then oscillations produce impact with surrounding molecules and energy is released in the environment or surrounding of matrix (Ganzler *et al.*, 1990; Sinquin *et al.*, 1993; Barnabas *et al.*, 1995; Onuska & Terry, 1995) and heating in this way is very fast. There is another thing to be noted that as larger is the dielectric constant of the solvents, there will be more and more heating effect (Jassie *et al.*, 1997). As a result microwaves cause heating of the whole matrix in lesser time. If the viscosity of the solvent is higher, it lowers the rotation of the molecules and as a result heating effect is reduced (Sinquin *et al.*, 1993; Camel & Bermond, 1999). The heating effect of microwave energy is strongly influenced by the nature of both the solvent and the solid plant matrix. Different types of solvents are used for microwave assisted extraction with a wide range of polarities, from n-hexane to water. Choice of solvent thus is of prime importance in microwave assisted extraction as sometimes the chosen solvent possesses a high dielectric constant and robustly takes up microwave energy, but not having the suitable solubility for the targeted molecule, in such situations a mixture of solvents can be used (Renoe, 1994).

Microwave assisted extraction is an emerging as well as striking alternative to conventional extraction methods such as maceration, Soxhlet, percolation, sonication, distillation and digestion. Now it has widely been used in laboratories as a digestion method for different sample types such as biological, geological and environmental matrixes. The heating effect of microwaves in lesser time and with lower requirement of solvent make it preferable for all solvents (Camel, 2000).

In present study, extraction of tea, coffee and Joshanda will be done by using simple boiling method on hot flame and by using microwave assisted extraction. Then a comparison for the amount of extract will be made for all of above mentioned products.

Aims & Objectives

The aim of this research work is to optimize a rapid & efficient method of extraction for a number of household products like tea, coffee and Joshanda.

LITERATURE REVIEW

The present research work is based on the extraction of the herbal products including different Tea samples, Peshawari Qehwa, Coffee and Joshanda which are still considered the mainstay of about 75 - 80% of the world population, mainly in the developing countries, for primary health care. This is primarily because of the general belief that herbal products are without any side effects besides being cheap and locally available.

Attard *et al.*, (2014) proved that microwave assisted extraction has been demonstrated as an efficient green technology for the recovery of D-limonene from orange waste. Microwave irradiation was shown to be a more efficient method when compared to conventional heating due to its high selectivity for D-limonene, significantly shortened extraction durations and D-limonene yielded twice that of conventional heating. Kinetic analysis of the extraction process indicated a typical two-step diffusion process, an initial stage of extraction from the exterior of the cells (1st stage) and diffusion of solute across the membrane (2nd stage). Diffusion coefficients for the initial stage of extraction from the exterior of cells (1st stage) for both conventional and microwave extraction demonstrated similar trends and activation energies. Interestingly, *trans*-membrane diffusion coefficient for the microwave assisted extraction at 110 °C was significantly high. Crucially, this was not observed with conventional heating suggesting that microwave radiation favorably interacts with the sample during extraction, causing simultaneous cell rupture and diffusion, resulting in greater yield. This provides an important insight into the development of extraction processes for orange peel.

Alupului *et al.*, (2012) carried out extensive research work whose aim was to increase extraction rate, yield and quality of products, associating microwave field with extraction of plant's active compounds such as flavonoids and phenolic acids from *Cynara scolymus* leaves. Compared to classical hot extraction methods, microwave extraction provided higher purities in active compounds while avoiding local over- exposure which might be very important assets for industry. Spectrophotometry was used to separate and quantify flavonoids and phenolic acids from herbal extracts of *Cynara scolymus*.

Orio *et al.*, (2012) compared hydrodistillation with microwave generated hydrodistillation in the extraction of several mint species cultivated in Piedmont: *Mentha spicata* L. var.*rubra, Mentha spicata* L. var. *viridis* and *Mentha piperita* L. Microwave generated hydrodistillation requires either fresh plant or rehydrated material was extremely fast and allows a reduction in energy consumption and overall cost. All the essential oils were analyzed by gas chromatography-mass spectrometry. A mechanism of microwave-generated essential oil extraction was thus proposed to explain the differences in the composition of the oil obtained from this environmentally friendly technique. The yields and composition percentages of the essential oils obtained by hydrodistillation and *in situ* microwave generated hydrodistillation of fresh and dried mint leaves lied in a relatively narrow range, although microwave generated hydrodistillation was faster. Microwave polarization effects and the water solubility of the components influence extract composition.

Guihua *et al.*, (2011) reported the extraction of aloe-emodin from Aloe by microwave assisted extraction. The effects of various factors including the solvent, the ratio (mL/g) of solvent to the sample, microwave irradiation time and microwave power, were discussed in the experiments. The yield of aloe-emodin was determined by HPLC. The optimized conditions for microwave assisted extraction of aloe-emodin were concluded as follows: the solvent was 80% ethanol (V/V) solution, microwave irradiation time was 3 minutes and microwave power was 340W. Additionally, HPLC finger print was developed for consistency evaluation of aloe. The similarities of 3 aloe samples obtained by microwave assisted extraction, ultrasound assisted extraction and soxhlet extraction were more than 0.9, indicating that three aloe samples were consistent. Compared with soxhlet extraction and ultrasound assisted extraction, microwave extraction was the rapid method with higher yield and less solvent consumption. Aloe samples treated by microwave assisted extraction, ultrasound assisted extraction and soxhlet extraction was observed using transmission electronic microscopy. The micrographs provided evidence of more breakage of chloroplast treated by microwave assisted extraction as compared to soxhlet and ultrasound assisted extraction.

Ushir *et al.,* (2011) observed different tea and coffee brand samples for their phenol, tannin & caffeine content. The product having highest phenol contents was Nescafe (5.03%). The products having least tannin contents were: Taj Mahal tea (3.99%), Mili tea (3.99%). Remaining product as: Wagh bakri, Red label, and Nescafe showed almost same percentage of tannins in range 4.50-5.50%. Among various tea & coffee brands, Bru had maximum quantity (9.65%) of tannin. Mili tea was containing highest quantity (1.72%) of caffeine among all the studied products.

Wakte *et al.*, (2011) studied a bio-active phyto-chemical, curcumin, which was isolated from dried rhizomes of *Curcuma longa* using Soxhlet, microwave, ultra-sonic and supercritical carbon dioxide assisted extraction techniques. The quantification of curcumin in resultant extracts was performed using pre-validated HPLC methodology. The critical parameters *viz.* effect of pre-irradiation and soaking solvent on the curcumin yield were studied. The extraction efficiency of all the above described techniques was established in terms of percent curcumin yields and extraction rate constants. Prior to extraction, microwave and ultra-sonic irradiation of dry curcuma powder resulted in 68.57 and 40.00% curcumin yield, respectively, whereas water soaked irradiated curcuma powder yielded 90.47 and 71.42% curcumin recovery respectively, during a total extraction period of five minutes. The maximum extraction rate constant of $47.49 \times 10^{-2} \, \text{min}^{-1}$ was observed when using microwave assisted acetone extract of water soaked curcuma rhizomes. The comparison of Soxhlet, microwave, ultra-sonic and supercritical carbon dioxide assisted extraction in terms of percent yield and required extraction period showed that microwave assisted extraction technique was more efficient for the curcumin extraction from powdered *C. longa* rhizomes.

Kodama *et al.*, (2010) analyzed and compared the tea samples in terms of phenolic contents and *in vitro* antioxidant capacity including tea bags, dehydrated leaves, and ready-to-drink preparations after standardization of the infusion preparation procedure. Total phenolics content in 1 cup of the different teas varied from 90 to 341 mg of catechin equivalents, and the highest and the lowest values were both those of the ready-to-drink products. Infusions prepared from tea bags had contents varying from 96 to 201 mg.200 mL^{-1}, and there were no significant differences among batches. The DPPH radical scavenging and the Oxygen Radical Absorbing Capacities (ORAC) varied largely among the different tea preparations, from 23 to 131 mmoles of Trolox Equivalents (TE).200 mL–1 (DPPH), and from 1.2 to 5.1 mmoles of TE.200 mL–1 (ORAC), but again there were no differences among infusions or ready-to-drink commercial preparations. However, the antioxidant capacity of ready-to-drink products was partially due to the presence of other non-phenolic compounds such as ascorbic acid.

Okoh et al., (2010) studied that Rosmarinus officinalis L. which is a perennial herb and belongs to the Lamiaceae family. Essential oils were obtained from this plant by hydrodistillation and solvent free microwave extraction. GC–MS analyses of the oils revealed the presence of 24 and 21 compounds in the essential oils obtained through hydrodistillation and solvent free microwave extraction, respectively. The total yield of the volatile fractions obtained through hydrodistillation and solvent free microwave extraction was 0.31% and 0.39%, respectively. Higher amounts of oxygenated monoterpenes such as borneol, camphor, terpene-4-ol, linalool, α-terpeneol (28.6%) were present in the oil of solvent free microwave extraction in comparison with hydrodistillation (26.98%). However, hydrodistillation oil contained more monoterpene hydrocarbons such as α-pinene, camphene, β-pinene, myrcene, α-phellanderene, 1,8-cineole, trans β-ocimene, γ-terpenene, and cis sabinene hydrate (32.95%) than solvent free microwave extraction extracted oil (25.77%). The essential oils obtained using the two methods of extraction were active against all the bacteria tested at a concentration of 10 mg ml^{-1}.

Srivastava *et al.,* (2009) studied the beneficial effects of tea (Camellia sinensis) leaves. The scientists are keen to explore the exact scientific reasons for such effects. The biochemical analysis of tea samples has indicated that they contain mostly polyphenols, mainly catechins. Catechins are present in food and beverages of plant origin. Keeping this in view, High Performance Liquid Chromatographic analysis was performed to estimate phenolic acids in 26 samples of tea (samples of Indian market brand, Indian tea garden and Nepal tea garden such as Janhit green tea™ (JGT), Tata tea premium™ (TT), Brooke bond Taaza™ (BBTZ), Lipton Darjeeling Tea™ (LDT), Tea city™ (TCT), Taj Mahal Tea™ (TAJM), Tetley Tea™ (TET), Assam dana (DASM), Assam dana 1 (ASMD-1), Assam dana 2 (ASMD-2), Assam Tea (ASM), North Bengal tea (NB), North Bengal dana-1 (DNB-1), North Bengal-2 (DNB-2), North Bengal Upper (NBUP), Darjeeling tea-FOP (DARJ-FOP), Darjeeling tea-FOP1 (DARJ-FOP1), Darjeeling tea- brake (DARJ-br), Indian tea garden (Darjeeling-ITGA (DARJ-ITGA), Darjeeling-ITGB (DARJ-ITGB), Darjeeling-ITGC (DARJ-ITGC) and Nepal tea garden (KP-33A, KP-16A, KP-9A, KP-C6 and KP-C2).

Anesini *et al.,* (2008) determined the total polyphenol content and the *in vitro* antioxidant capacity of green and black tea, cultivated and industrialized in Argentina. Twelve samples of eight brands were analyzed. The total polyphenol content was determined according to the International Organization for Standardization method (ISO) 14502-1 for the determination of substances characteristic of green and black tea. The antioxidant capacity was determined by the ferric thiocyanate method (FTC) and the 1,1-diphenyl-2-picrylhydrazyl (DPPH) free-radical scavenging assay. Green tea showed a higher polyphenol content than black tea. The total polyphenol concentration in green tea was found to vary from 21.02 (1.54 to 14.32 (0.45% of gallic acid equivalents (GAE), whereas in black tea, the polyphenol content ranged from 17.62 (0.42 to 8.42 (0,55% of GAE ($P < 0.05$). A similar profile was observed for the antioxidant capacity determined by both methods. The antioxidant activities were well correlated with the total polyphenol content ($r 2$) 0.9935 for the ferric thiocyanate method and r2) 0.9141 for the 1,1-diphenyl-2-picrylhydrazyl free-radical scavenging assay).

Bayramoglu and Serplis (2008) examined the applicability of solvent-free microwave extraction in the extraction of essential oil from *Origanum vulgare* L. and the effects of microwave power and extraction time on the yield and composition of the product were investigated. Specific gravity and refractive index of the essential oil and its solubility in alcohol were also examined. Hydrodistillation was performed as control. GC–MS/FID was used for the determination and quantification of aroma compounds in the essential oils. Solvent free microwave extraction offered significantly higher essential oil yields (0.054 mL/g) as compared to hydrodistillation (0.048 mL/g). When 622 W microwave powers were used in solvent free microwave extraction, conventional process time was reduced by 80%. The main aroma compound of oregano essential oil was found to be thymol (650–750 mg/mL). No significant differences were obtained in the compositions and physical properties of oregano essential oils obtained by solvent free microwave extraction and hydrodistillation.

Chen *et al.*, (2008) introduced an advanced apparatus for extraction of flavonoids from *Herba Epimedii* which utilized a microwave technique for heating in a dynamic mode. A TM_{010} microwave resonance cavity was applied to concentrate the microwave energy; the power needed to achieve efficient extraction pressurized microwave extraction Ultrasonic extraction, Heat reflux extraction and Soxhlet extraction. The four flavonoids in extract were identified by HPLC-DAD spectra in the different extraction methods. The results showed that the extraction yield of flavonoids in *Herba Epimedii* obtained by dynamic microwave extraction was higher than that obtained by ultrasonic extraction, reflux extraction and soxhlet and extraction. It was concluded that chemical components would not decompose easily when increasing the extraction time in dynamic microwave assisted extraction. This was the benefit of dynamic extraction system which could transfer the analytes extracted from the sample matrix and flowed fresh solvent to the out of extraction vessel.

Golmakani & Rezaei (2008) observed through experiments that microwave-assisted hydrodistillation was an advanced hydrodistillation technique utilizing a microwave oven in the extraction process. Microwave assisted hydrodistillation of essential oils from the aerial parts (tops) of *Thymus vulgaris* L. (common thyme) was studied and the results were compared with those of the conventional hydrodistillation in terms of extraction time, extraction yield/efficiency, chemical composition, quality of the essential oils and cost of the operation. Microwave assisted hydrodistillation was superior in terms of saving energy and extraction time (75 min, compared to 4 h in hydrodistillation). Scanning electron microscopy of thyme leaves undergone hydrodistillation and microwave assisted hydrodistillation provided evidences as to a sudden rupture of essential oil glands with microwave hydro distillation. Gas chromatography–mass spectrometry analysis of the extracted essential oils indicated that the use of microwave irradiation did not adversely influence the composition of the essential oils. Microwave assisted hydrodistillation was found to be a green technology.

Lianfu & Zelang (2008) studied and compared the extracting technology including ultrasonic and microwave assisted extraction and ultrasonic assisted extraction of lycopene from tomato paste. The results showed that the optimal conditions for ultrasonic microwave assisted extraction were 98 W microwave power together with 40 KHz ultrasonic processing, the ratio of solvents to tomato paste was 10.6:1 (V/W) and the extracting time should was 367 s; as for ultrasonic assisted extraction, the extracting temperature was 86.4 °C, the ratio of the solvents to tomato paste was 8.0:1 (V/W) and the extracting time was 29.1 min, while the percentage of lycopene yield was 97.4% and 89.4% for ultrasonic microwave assisted extraction and ultrasonic assisted extraction, respectively. These results implied that ultrasonic microwave assisted extraction was far more efficient extracting method than ultrasonic assisted extraction.

Turkmen *et al.,* (2006) studied the effect of the use of water and different organic solvents such as acetone, *N,N*-dimethylformamide (DMF), ethanol or methanol at various concentrations on the total polyphenol content and antioxidant activity was studied for the black tea and mate tea. Polyphenol contents of extracts were determined using ferrous tartarate (method # 1) and Folin–Ciocalteu (method # 2) assays. For black tea, 50% DMF extract showed the highest polyphenol content of 131.9 mg/g and 99.8 mg GAE/g by method # 1 and method # 2, respectively. For mate tea, 50% acetone showed the highest polyphenol content of 132.5 mg/g and 120.4 mg GAE/g by method # 1 and method # 2, respectively. Fifty percent ethanol extract from mate tea and 50% acetone from black tea had the greatest antioxidant activity. The results showed that solvent with different polarity had significant effect on polyphenol content and antioxidant activity. A high correlation between polyphenol content and antioxidant activity of tea extracts was observed.

Yao *et al.*, (2006) observed the phenolic compounds of tea water extract. Thea flavins (TF), thea rubigins (TR) and thea brownins (TB) were the major polyphenols that determine the quality of black tea. These compounds were measured in 56 leaf teas and teabags sampled from Australian supermarkets in Queensland. The various quantities of TF, ranging from 0.29% to 1.25%, indicate a quality difference that exists among the teas studied. Low TF content in black tea may be due to over-fermenting and/or long periods of storage. The solubility of TR and TB from teabags ranged from 82% to 92%, indicating that the permeability of teabags was variable. Variable quantities of TF in Australian teas show instability and a tendency of TF to oxidize during storage. Total polyphenols in green teas ranged from 14% to 34%, indicating a large variation, which was not reflected in price. The solubility of total polyphenols from teabags has been proposed as a useful quality index of the filtering paper used for the teabags. This chemical analysis of phenolic compounds in commercial teas may be a potential tool for the quality control of Australian manufactured and imported teas in Australian markets.

Atoui *et al.*, (2005) studied different tea and herbal infusions for their polyphenolic content, antioxidant activity and phenolic profile. The total phenolics recovered by ethyl acetate from the water extract, were determined by the Folin–Ciocalteu procedure and ranged from 88.1 ± 0.42 (Greek mountain tea) to 1216 ± 32.0 mg (Chinese green tea) GAE (Gallic acid equivalents)/cup. The antioxidant activity was evaluated by two methods, DPPH and chemiluminescence assays, using Trolox and quercetin as standards. The EC_{50} of herbal extracts ranged from 0.151 ± 0.002 mg extract/mg DPPH (0.38

quercetin equivalents and 0.57 Trolox equivalents), for Chinese green tea, to 0.77 ± 0.012 mg extract/mg DPPH (0.08 quercetin equivalents and 0.13 Trolox equivalents), for Greek mountain tea. Chemi-luminescence assay results showed that the IC_{50} ranged from $0.17 \pm 3.4 \times 10^{-3}$ µg extract/ml of the final solution in the measuring cell (1.89 quercetin and 5.89 Trolox equivalents) for Chinese green tea, to $1.10 \pm 1.86 \times 10^{-2}$ g extract/ml of the final solution in the measuring cell (0.29 quercetin and 0.90 Trolox equivalents) for Greek mountain tea. The phenolic profile in the herbal infusions was investigated by LC-DAD-MS in the positive electrospray ionization (ESI^{+}) mode. About 60 different flavonoids, phenolic acids and their derivatives were identified.

Lucchesi *et al.*, (2004) defined that Solvent-free microwave extraction was a combination of microwave heating and dry distillation, performed at atmospheric pressure without added any solvent or water. Isolation and concentration of volatile compounds were performed by a single stage. Solvent free microwave extraction was compared with a conventional technique, hydro-distillation, for the extraction of essential oil from three aromatic herbs: basil (*Ocimum basilicum* L.), garden mint (*Mentha crispa* L.), and thyme (*Thymus vulgaris* L.). The essential oils extracted by solvent free microwave extraction for 30min were quantitatively (yield) and qualitatively (aromatic profile) similar to those obtained by conventional hydro-distillation for 4.5h. The solvent free microwave extraction method yielded an essential oil with higher amounts of more valuable oxygenated compounds, and allowed substantial savings of costs, in terms of time, energy and plant material. Solvent free microwave extraction was proved to be green technology and appeared as a good alternative for the extraction of essential oils from aromatic plants.

Liang *et al.*, (2003) assessed the chemical composition, color differences of black tea infusions and their relationships with sensory quality by tea tasters. There were significant correlations between the individual quality attributes. Content of caffeine, nitrogen, amino acids, polyphenols, gallocatechin (GC), epigallocatechin (EGC), catechin

(C), epicatechin (EC), epicatechin gallate (ECG), catechin gallate (CG), total catechins, theaflavin (TF) and theaflavin-3'-gallate (TF3'G) and infusion colour indicators of ΔL, Δa, Δb and ΔE were significantly correlated to total quality score (TQS). The parameters correlating significantly with the TQS were classified into four groups. Group 1 was compounds containing nitrogen, group 2 phenol compounds, group 3 tea pigments and group 4 infusion colour indicators. Four principal components were screened from the four groups as independent variables for constructing regression equations for estimation of black tea quality by principal component analysis. The regression of the TQS upon the principal components gave a highly significant relationship.

Pan *et al.*, (2003) described a microwave-assisted extraction method for the extraction of tea polyphenol and tea caffeine from green tea leaves. Various experimental conditions, such as ethanol concentration (0/100%, v/v), microwave assisted extraction time (0.5/8 min), liquid/solid ratio (10:1/25:1 ml g1), pre-leaching time (0/90 min) before MAE and different solvents for the microwave assisted extraction procedure were investigated to optimize the extraction. The extraction of tea polyphenols and tea caffeine with microwave assisted extraction for 4 min (30 and 4%) were higher than those of extraction at room temperature for 20 h, ultrasonic extraction for 90 min and heat reflux extraction for 45 min (28 and 3.6%), respectively. From the points of extraction time, the extraction efficiency and the percentages of tea polyphenols or tea caffeine in extracts, microwave assisted extraction was more effective than the conventional extraction methods studied.

Pan *et al.*, (2002) used different extraction techniques like microwave-assisted extraction, extraction at room temperature, heat reflux extraction, ultra- sonic extraction and soxhlet extraction for the extraction of tanshinones (Cryptotanshinone, Tanshinone I and Tanshinone IIA) from *Salvia miltiorrhiza* bunge. The extracts were analyzed by high performance liquid chromatography without any treatment. The results showed that the percentage extraction of Cryptotanshinone, Tanshinone I and Tanshinone IIA from *S. miltiorrhiza* bunge by microwave assisted extraction was equivalent with and in fact higher than that of conventional extraction methods. Microwave assisted extraction only needed 2 min, whereas extraction at room temperature, heat reflux extraction, and ultrasonic extraction and soxhlet extraction needed 24 h, 45, 75 and 90 min, respectively. Due to the considerable saving of time and high extraction efficiency, microwave assisted extraction was more effective than the conventional methods.

Dai *et al.*, (2001) studied the use of the microwave-assisted process for the extraction of azadirachtin related limonoids from various parts of the neem tree under different operating conditions. The influence of microwave power, solvent, and irradiation time on the recovery of azadirachtin related limonoids was studied. The efficiency of the microwave-assisted extraction of the seed kernel, the seed shell, the leaf, and the leaf stem was compared to that of conventional extraction methods. The content of azadirachtin related limonoids in the extracts was estimated with a vanillin-based colorimetric assay and a multivariate calibration technique. The results showed that the microwave assisted extraction technique can enhance the extraction of azadirachtin related limonoids from different parts of neem possessing microstructures. Investigation of the influence of the solvent also indicated that the solvent used not only influenced the efficiency but also affected the selectivity of the microwave assisted extraction.

MATERIALS AND METHODS

The present research work was carried in Biochemistry Lab (Dept of Botany) in Lahore College for Women University Lahore. Samples were collected from the nearby local market.

3.1:   Samples

10 different samples were taken, which were tea samples (Green tea, Vital tea, Supreme tea, Danedar tea, Tez Dam tea, Tetley and Lipton tea), Qehwa (Peshawari Qehwa), Coffee (Nescafe Classic) and Joshanda.

3.2:   Sample preparation

For each extraction 5g of the samples were weighed accurately and placed in the beakers. Measured quantity of distilled H_2O (50ml) was added in beaker as an extracting solvent along with the plant material and then heated for extraction. Two different methods i.e., microwave heating and stove heating were then opted to extract the bioactive compounds from plant material.

3.3:   Microwave Assisted Extraction

For microwave assisted extraction, each sample was heated in the microwave at full power level, described 1000W (Model HGN-45100EB) for 2 minutes for each sample each time. For Microwave assisted extraction, each plant sample was stained after heating by using common household Stainer and poured into pre-weighed glass vials and was subjected to shade drying.

3.4: Stove Heating

For obtaining the extract by stove heating extraction, each sample was heated at normal heat of stove for 2 minutes for each sample each time. For stove heating extraction, each plant sample was stained after heating by using common household stainer and poured into pre-weighed glass vials and was subjected to shade drying.

3.5: Quantification of extract

After drying, extracts were weighed by the formula

Weight of extract = (wt of glass vial + wt of extract) – wt of glass vial

Then % age of the extract was found by applying the formula

% age of extract = wt of extract/ wt of the sample * 100

3.6: Qualitative Tests:

a) Alkaloids

2ml of Dragon droff reagent was added in 2ml of filtrate. Turbid orange coloration appeared as end point.

b)   Phenolics

2ml of 10% led acetate is added in 1ml of filtrate. Brown ppt formed as end point.

c)   Flavonoids

In 1ml of filtrate solution, 2ml of NaOH was added. Golden yellow color appeared as end point.

d)   Steroids

In 1ml of filtrate, 10ml of Chloroform and 10ml of H_2SO_4 (slowly by side of test tube) were added. As a result, the upper layer showed red colour and H_2SO_4 layer gave yellow/green coloration.

e)   Terpenoids

2ml of Chloroform and few drops of H_2SO_4 were added in 1ml of filtrate. The results gave reddish brown interface.

3.7: Quantitative Tests

a)   Phenolics

To the 1 ml of sample (1mg/ml), 1 ml of Folin and Ciocalteu's (FC reagent) phenol reagent was added. After 3 minutes saturated Na_2CO_3 (1ml) was added. Then distilled water was added to make total volume 10ml with distilled water. The solution was set aside in dark for about 90 minutes. Absorbance was measured at 725nm on UV-visible spectrophotometer. Gallic acid was usedas a standard phenolics reagent for constructing the standard curve passing through tha same chemical treatment (200-100µg/mL).

b)   Flavonoids

250µL extract (1mg/ml) was added with 1.25ml distilled water and 75µL of 5% sodium nitrite ($NaNO_2$). After five minutes, 150µL of 10% (Aluminium chloride) $AlCl_3$ were added. After six minutes of this addition, 500µL of 1M NaOH and 275µL of distilled water were added. Then the solutions were mixed well. Absorbance was measured at 510nm. Catechin was used to construct the standard curve (20-120µg/ml).

3.8: Statistical analysis

Statistical analysis was done by comparing the means by applying the ANOVA (Analysis of variance) and each value was a mean of five replicates with standard deviation (mean ± SD). Means represented in the same column not having a common superscript, differ significantly ($P<0.05$) according to Duncan's new multiple range test.

RESULTS AND DISCUSSION

4.1: Extraction of Qehwa:

Table 4.1 shows the microwave assisted extraction of Qehwa sample as well as its boiling on stove. Microwave assisted extraction was done at 1000W of power level for 2 minutes while stove heating was done for 2 minutes in an open pan and water was used as a solvent in both cases.

It is clear from the table that there is no significant difference between the Microwaves assisted extraction and stove heating extraction of Qehwa. An average of 1.554g of extract was obtained from Microwave assisted extraction and an average of 1.334g of extract was obtained from stove heating extraction.

Table 4.1: Microwave assisted extraction and Stove Heating of Qehwa.

Method of Extraction	Serial Number	Weight of extract	Mean Value
Microwave assisted extraction	1	1.67g	
	2	1.39g	
	3	1.67g	
	4	1.31g	
	5	1.73g	$1.554^{a} \pm 0.07$
Stove heating	1	1.42g	
	2	1.58g	$1.334^{a} \pm 0.10$
	3	1.49g	
	4	1.25g	
	5	0.93g	

Each value presented is the mean of 5 replicate with standard deviation (mean ± SD). Means which don't have a common superscript in same column differ significantly (P<0.05) by Duncan's new multiple range test.

4.2: Extraction of Green Tea:

Table 4.2 explains the microwave assisted extraction of Green tea sample and it's boiling on stove. Microwave assisted extraction was done at 1000W of power level for 2 minutes while stove heating was done for 2 minutes in an open pan and water was used as a solvent in both cases.

It is obvious from the table that there is a significant difference between the Microwave assisted extraction and stove heating extraction of Green tea. An average of 1.27g of extract was obtained from Microwave assisted extraction and an average of 1.002g of extract was obtained from stove heating extraction. It is clear from the table that microwave assisted extraction gives the higher value of the extract as compared to stove heating extraction.

Table 4.2: Microwave assisted extraction and Stove Heating of Green Tea.

Serial Number	Method of Extraction	Weight of extract	Mean Value
1	Microwave assisted extraction	1.51g	$1.27^{\,a} \pm 0.061644$
2		1.28g	
3		1.09g	
4		1.27g	
5		1.2g	
6	Heating on Stove	1.04g	$1.002^{\,b} \pm 0.02932$
7		0.95g	
8		1.05g	
9		1.07g	
10		0.9g	

Each value presented is the mean of 5 replicate with standard deviation (mean ± SD). Means which don't have a common superscript in same column differ significantly (P<0.05) by Duncan's new multiple range test.

4.3: Extraction of Joshanda:

Table 4.3 gives the Microwave assisted extraction of Joshanda sample along with its boiling on stove. Microwave assisted extraction was done at 1000W of power level for 2 minutes while stove heating was done for 2 minutes in an open pan and water was used as solvent in both cases.

It is apparent from the table that there is a significant difference between the Microwave assisted extraction and stove heating extraction of Joshanda. An average of 3.76g extract was obtained from Microwave assisted extraction and an average of 3.12g of extract was obtained from stove heating extraction.

The quantity of extract obtained from the microwave assisted extraction of Joshanda is considerably higher than that of the stove heating extraction.

Table 4.3: Microwave assisted extraction & Stove Heating of Joshanda.

Serial Number	Method of Extraction	Weight of extract	Mean Value
1	Microwave assisted extraction	3.7g	3.76 ^{a} ± 0.077974
2		3.9g	
3		4g	
4		3.7g	
5		3.5g	
6	Heating on Stove	3g	3.12 ^{b} ± 0.065727
7		3.2g	
8		3.3g	
9		3.2g	
10		2.9g	

Each value presented is the mean of 5 replicate with standard deviation (mean ± SD). Means which don't have a common superscript in same coloumn differ significantly (P<0.05) by Duncan's new multiple range test.

4.4: Extraction of Coffee:

Table 4.4 reveals the Microwave assisted extraction of Coffee sample as well as its boiling on stove. Microwave assisted extraction was done at 1000W of power level for 2 minutes while stove heating was done for 2 minutes in an open pan and water was used as solvent in both cases.

It is clear that there is a significant difference between the Microwave assisted extraction and stove heating extraction of Coffee. An average of 3.76g extract was obtained from Microwave assisted extraction and an average of 3.14g of extract was obtained from stove heating extraction.

The results obtained by microwave assisted extraction are more advantageous because it gives the higher quantity of extract than the stove heating extraction.

Table 4.4: Microwave assisted extraction & Stove Heating of Coffee.

Serial Number	Method of Extraction	Weight of extract	Mean Value
1		4.1g	
2	Microwave assisted extraction	3.7g	
3		3.4g	3.76 [a] ± 0.104307
4		3.9g	
5		3.7	
6		2.8g	
7	Heating on Stove	3.4g	
8		3.1g	3.14 [b] ± 0.092087
9		3.1g	
10		3.3g	

Each value presented is the mean of 5 replicate with standard deviation (mean ± SD). Means which don't have a common superscript in same coloumn differ significantly ($P<0.05$) by Duncan's new multiple range test.

4.5: Extraction of Tapal Tez Dam Tea:

Table 4.5 estimates the Microwave assisted extraction of Tapal Tez Dam Tea sample and its boiling on stove as well. Microwave assisted extraction was done at 1000W of power level for 2 minutes while stove heating was done for 2 minutes in an open pan and water was used as solvent in both cases.

It is evident from the table that there is a significant difference between the Microwave assisted extraction and stove heating extraction of Tapal Tez Dam Tea. An average of 1.052g extract was obtained from Microwave assisted extraction and an average of 0.64g of extract was obtained from stove heating extraction.

The microwave assisted extraction gives higher rate of extraction as compared to the stove heating extraction method.

Table 4.5: Microwave assisted extraction & Stove Heating of Tapal Tez Dam Tea.

Serial Number	Method of Extraction	Weight of extract	Mean Value
1		1.11g	
2	Microwave assisted extraction	1.06g	
3		1.03g	$1.052^a \pm 0.01397$
4		1.03g	
5		1.03g	
6		0.6g	
7	Heating on Stove	0.73g	
8		0.55g	$0.64^b \pm 0.032125$
9		0.6g	
10		0.72g	

Each value presented is the mean of 5 replicate with standard deviation (mean ± SD). Means which don't have a common superscript in same coloumn differ significantly (P<0.05) by Duncan's new multiple range test.

4.6: Extraction of Lipton Tea:

Table 4.6 describes the Microwave assisted extraction of Lipton Tea sample and also it's boiling on stove. Microwave assisted extraction was done at 1000W of power level for 2 minutes while stove heating was done for 2 minutes in an open pan and water was used as solvent in both cases.

It is exposed from the table that there is no significant difference between the Microwave assisted extraction and stove heating extraction of Lipton Tea. An average of 1.26g extract was obtained from Microwave assisted extraction and an average of 0.56g of extract was obtained from stove heating extraction.

Table 4.6: Microwave assisted extraction & Stove Heating of Lipton Tea.

Serial Number	Method of Extraction	Weight of extract	Mean Value
1	Microwave assisted extraction	1.1g	$1.26^{a} \pm 0.258612$
2		1g	
3		0.8g	
4		1g	
5		2.4g	
6	Heating on Stove	0.9g	$0.56^{a} \pm 0.092087$
7		0.4g	
8		0.3g	
9		0.6g	
10		0.6g	

Each value presented is the mean of 5 replicate with standard deviation (mean ± SD). Means which don't have a common superscript in same coloumn differ significantly (P<0.05) by Duncan's new multiple range test.

4.7: Extraction of Vital Tea:

Table 4.7 is distinguishable for the Microwave assisted extraction and boiling on stove of Vital Tea sample. Microwave assisted extraction was done at 1000W of power level for 2 minutes while stove heating was done for 2 minutes in an open pan and water was used as solvent in both cases.

It is explicit from the table that there is no significant difference between the Microwave assisted extraction and stove heating extraction of Lipton Tea. An average of 0.92g extract was obtained from Microwave assisted extraction and an average of 0.84g of extract was obtained from stove heating extraction.

Table 4.7: Microwave assisted extraction & Stove Heating of Vital Tea.

Serial Number	Method of Extraction	Weight of extract	Mean Value
1	Microwave assisted extraction	0.8g	0.92 [a] ± 0.33466
2		1g	
3		0.9g	
A~4		0.9g	
5		1g	
6	Heating on Stove	0.5g	0.84 [a] ± 0.143108
7		0.8g	
8		1.1g	
9		0.5g	
10		1.3g	

Each value presented is the mean of 5 replicate with standard deviation (mean ± SD). Means which don't have a common superscript in same coloumn differ significantly (P<0.05) by Duncan's new multiple range test.

4.8: Extraction of Tetley Tea:

Table 4.8 is self explanatory for the Microwave assisted extraction and boiling on stove of Tetley Tea sample. Microwave assisted extraction was done at 1000W of power level for 2 minutes while stove heating was done for 2 minutes in an open pan and water was used as solvent in both cases.

It is prominent from the table that there is a significant difference between the Microwave assisted extraction and stove heating extraction of Tetley Tea. An average of 0.8g extract was obtained from Microwave assisted extraction and an average of 0.52 of extract was obtained from stove heating extraction.

The extract of Tetley Tea obtained by the microwave assisted extraction is higher than stove heating extraction.

Table 4.8: Microwave assisted extraction & Stove Heating of Tetley Tea.

Serial Number	Method of Extraction	Weight of extract	Mean Value
1	Microwave assisted extraction	0.9g	$0.8\ ^a \pm 0.56569$
2		0.6g	
3		0.7g	
4		0.9g	
5		0.9g	
6	Heating on Stove	0.4g	$0.52\ ^b \pm 0.033466$
7		0.6g	
8		0.5g	
9		0.6g	
10		0.5g	

Each value presented is the mean of 5 replicate with standard deviation (mean ± SD). Means which don't have a common superscript in same coloumn differ significantly (P<0.05) by Duncan's new multiple range test.

4.9: Extraction of Supreme Tea:

Table 4.9 is observable for the Microwave assisted extraction and boiling on stove of Supreme Tea sample. Microwave assisted extraction was done at 1000W of power level for 2 minutes while stove heating was done for 2 minutes in an open pan and water was used as solvent in both cases.

It is precise from the table that there is a significant difference between the Microwave assisted extraction and stove heating extraction of Supreme Tea. An average of 0.84g extract was obtained from Microwave assisted extraction and an average of 0.48g of extract was obtained from stove heating extraction. Microwave assisted extraction gives the best results of extraction than the stove heating extraction.

Table 4.9: Microwave assisted extraction & Stove Heating of Supreme Tea.

Serial Number	Method of Extraction	Weight of extract	Mean Value
1	Microwave assisted extraction	0.7g	0.84 [a] ± 0.045607
2		1g	
3		0.9g	
4		0.8g	
5		0.8g	
6	Heating on Stove	0.5g	0.48 [b] ± 0.017889
7		0.5g	
8		0.5g	
9		0.4g	
10		0.5g	

Each value presented is the mean of 5 replicate with standard deviation (mean ± SD). Means which don't have a common superscript in same coloumn differ significantly (P<0.05) by Duncan's new multiple range test.

4.10: Extraction of Tapal Danedar Tea:

Table 4.10 is conclusive for the Microwave assisted extraction and boiling on stove of Tapal Danedar Tea sample. Microwave assisted extraction was done at 1000W of power level for 2 minutes while stove heating was done for 2 minutes in an open pan and water was used as solvent in both cases.

It is accessible from the table that there is a significant difference between the Microwave assisted extraction and stove heating extraction of Tapal Danedar Tea. An average of 0.86g extract was obtained from Microwave assisted extraction and an average of 0.44g of extract was obtained from stove heating extraction.

Microwave assisted extraction gives the higher value of extraction while the stove heating gives the minimum value of the extract.

Table 4.10: Microwave assisted extraction & Stove Heating of Tapal Danedar.

Serial Number	Method of Extraction	Weight of extract	Mean Value
1	Microwave assisted extraction	0.7g	0.86 [a] ± 0.045607
2		0.9g	
3		0.9g	
4		0.8g	
5		1g	
6	Heating on Stove	0.6g	0.44 [b] ± 0.035777
7		0.4g	
8		0.4g	
9		0.4g	
10		0.4g	

Each value presented is the mean of 5 replicate with standard deviation (mean ± SD). Means which don't have a common superscript in same column differ significantly ($P<0.05$) by Duncan's new multiple range test.

4.11: Qualitative and quantitative analysis of microwave extracted and stove extracted plant material.

Table 4.11 a & b tells about the phyto-chemical estimation of leaves' extract of all the plant samples involved. It also showed that microwave assisted extraction of plant samples showed higher amounts of alkaloids, phenolics, flavonoids, tannins and steroids as compared to stove heated extract.

Table 4.12 shows the effect of microwave assisted extraction and stove heating on the amount of Phenolics present in the extract in comparison to each other as well as to the whole samples.

It shows that there is a significant difference of amount of Phenolics from microwave assisted extract and stove heated extract of each sample. In Qehwa sample, for example, by microwave extraction amount of Phenolics present was 40 µg equivalent of gallic acid while in stove heated sample it was only 28.96 µg equivalent of gallic acid.

In case of Green Tea sample, there is a clear difference of amount of Phenolics from microwave assisted extract and stove heated extract. By microwave extraction, the amount of Phenolics present was 45.90 µg equivalent of gallic acid while in stove heated extract it was 34.00 µg.

The sample of Tapal Tez Dam Tea also shows noticeable difference of amount of Phenolics from microwave assisted extract and stove heated extract. The amount of Phenolics estimated by microwave extraction was 88.33 µg equivalent of gallic acid while in stove heated extract it was 61.66 µg.

Similarly all the other samples show different amount of Phenolics in both microwave extract and stove heated extract. All the samples including Joshanda, Coffee, Lipton Tea, Vital Tea, Tetley Tea, Supreme Tea and Tapal Danedar Tea showed high amount of Phenolics when treated with microwave extraction while in case of Stove heated extraction, the amount of Phenolics was lesser. In Tapal Tez Dam Tea sample, maximum amount of Phenolics by microwave assisted extraction was 88.33 µg equivalent of gallic acid. On the other hand, the sample of Tapal Danedar Tea sample when treated with stove heating extraction, the highest amount of Phenolics was 71.00 µg.

So it is clear from the results that microwave assisted extraction gives higher amount of Phenolics than stove heating extraction.

Table 4.13 shows the effect of microwave assisted extraction and stove heating on the amount of Flavonoids present in the extract in comparison to each other as well as to the whole samples.

It shows that there is a significant difference of amount of Flavonoids from microwave assisted extract and stove heated extract of each sample. In Qehwa sample, for example, by microwave extraction amount of Phenolics present was 92.00 µg equivalent of Catechin while in stove heated sample it was only 80.33 µg.

The Coffee Sample shows that there is a significant difference between the amounts of Flavonoids estimated from microwave extract and stove heated extract. The amount of Flavonoids obtained by microwave assisted extraction was 88.33 µg equivalent of Catechin while in stove heated extract it was 69.33 µg.

The sample of Lipton Tea sample shows distinct difference between the amounts of Flavonoids obtained from microwave extract and stove heated extract. The amount of Flavonoids by microwave assisted extraction was 91.33 µg equivalent of Catechin in stove heated extract it was 71.93 µg.

The other samples also show different amounts of Flavonoids by microwave extraction and stove heating extraction. All the samples used which are named as Green Tea, Joshanda, Tapal Tez Dam Tea, Vital Tea, Tetley Tea, Supreme Tea and Tapal Danedar Tea showed higher amounts of Flavonoids when treated with microwave assisted extraction method while in case of stove heating extraction, the amount of Flavonoids was lesser. The highest amount of Flavonoids by microwave assisted extraction was noticed in Supreme Tea sample. In case of stove heating extraction, the sample of Vital Tea showed the highest amount of Flavonoids which was 86.66 µg.

The results show that the microwave assisted extraction method gives higher amounts of Flavonoids than Stove heating extraction.

Table 4.11(a): Qualitative phyto–chemical screening of microwave extracted and stove extracted plant material.

Serial Number	Plant Material	Sample	Alkaloids	Phenolics	Flavonoids	Steroids	Terpenoids
1	Qehwa	MAE	+++	++++	++	--	--
		Stove extraction	++	++	++	--	--
2	Green Tea	MAE	+++	++++	+++		++
		Stove Extraction	++	++	++		
3	Joshanda	MAE	++	+++	+++		++
		Stove extraction	++	+			+ -
4	Coffee	MAE	+++++	+++	+++		++
		Stove Extraction	+++	++	++	+ -	
5	Tapal Tez Dam Tea	MAE	+++	+++	++		++
		Stove Extraction	+	++	++		

Table 4.11(b): Qualitative phyto–chemical screening of microwave extracted and stove extracted plant material.

45

Serial Number	Plant Material	Sample	Alkaloids	Phenolics	Flavonoids	Steroids	Terpenoids
6	Lipton Tea	MAE	+++	+++	+++		+++
		Stove extraction	++	++	+++		+
7	Vital Tea	MAE	+++	+++	++		++
		Stove Extraction	++	++	++		
8	Tetley Tea	MAE	+++	+++	++		++
		Stove extraction	++	++	++		
9	Supreme Tea	MAE	+++	+++	+++		++
		Stove Extraction	+++	++	+++		
10	Tapal Danedar Tea	MAE	++	++	++		++
		Stove Extraction	++	++	++		

Table 4.12: Quantitative estimation of Phenolics from microwave extracted and stove extracted plant material.

Serial Number	Plant Material	Sample	Mean Value
1	Qehwa	MAE	$40.00^{j} \pm 1.20$
		Stove Heating Extraction	$28.96^{lm} \pm 0.55$
2	Green Tea	MAE	$45.90^{i} \pm 0.75$
		Stove Heating Extraction	$34.00^{k} \pm 0.50$
3	Joshanda	MAE	$30.83^{l} \pm 1.60$
		Stove Heating Extraction	$27.16^{m} \pm 0.76$
4	Coffee	MAE	$63.33^{f} \pm 1.15$
		Stove Heating Extraction	$57.83^{h} \pm 0.76$
5	Tapal Tez Dam Tea	MAE	$88.33^{a} \pm 1.52$
		Stove Heating Extraction	$61.66^{g} \pm 1.52$
6	Lipton Tea	MAE	$80.33^{c} \pm 1.52$
		Stove Heating Extraction	$70.00^{e} \pm 1.00$
7	Vital Tea	MAE	$79.00^{d} \pm 1.00$
		Stove Heating Extraction	$60.96^{fg} \pm 1.76^{(}$
8	Tetley Tea	MAE	$78.33^{d} \pm 1.52$
		Stove Heating Extraction	$70.00^{e} \pm 1.73$
9	Supreme Tea	MAE	$85.33^{b} \pm 1.52$
		Stove Heating Extraction	$70.66^{e} \pm 1.52$
10	Tapal Danedar Tea	MAE	$83.66^{b} \pm 1.52$
		Stove Heating Extraction	$71.00^{e} \pm 1.00$

Each value presented is the mean of 3 replicate with standard deviation (mean ± SD). Means which don't have a common superscript in same column differ significantly ($P<0.05$) by Duncan's new multiple range test.

Table 4.13: Quantitative estimation of Flavonoids from microwave extracted and stove extracted plant material.

Serial Number	Plant Material	Sample	Mean Value
1	Qehwa	MAE	$92.00^{ab} \pm 1.07$
		Stove Heating Extraction	$80.33^{e} \pm 1.15$
2	Green Tea	MAE	$80.66^{e} \pm 1.52$
		Stove Heating Extraction	$72.66^{h} \pm 1.52$
3	Joshanda	MAE	$41.00^{k} \pm 1.00$
		Stove Heating Extraction	$31.03^{l} \pm 1.67$
4	Coffee	MAE	$88.33^{cd} \pm 1.52$
		Stove Heating Extraction	$69.33^{j} \pm 1.52$
5	Tapal Tez Dam Tea	MAE	$88.66^{c} \pm 1.52$
		Stove Heating Extraction	$70.33^{ij} \pm 1.52$
6	Lipton Tea	MAE	$91.33^{b} \pm 0.57$
		Stove Heating Extraction	$71.93^{hi} \pm 1.00$
7	Vital Tea	MAE	$86.66^{d} \pm 1.52$
		Stove Heating Extraction	$70.00^{ij} \pm 1.73$
8	Tetley Tea	MAE	$81.33^{e} \pm 1.52$
		Stove Heating Extraction	$78.66^{f} \pm 1.52$
9	Supreme Tea	MAE	$93.33^{a} \pm 1.54$
		Stove Heating Extraction	$75.33^{g} \pm 1.52$
10	Tapal Danedar Tea	MAE	$88.33^{cd} \pm 1.52$
		Stove Heating Extraction	$69.00^{j} \pm 1.00$

Each value presented is the mean of 5 replicate with standard deviation (mean ± SD). Means which don't have a common superscript in same column differ significantly (P<0.05) by Duncan's new multiple range test.

DISCUSSION

In Pharmaceutical sciences, extraction is defined as the separation of bioactive components of plant from the inactive components with the help of solvents of specific polarity. The products of these extraction procedures are usually in the form of some liquids, semisolids or powders used only for oral or external use. The main purpose of standardized extraction procedures is to make them suitable for human use as drugs and to eradicate the inert or unwanted material by treatment with a selective solvent (Handa *et al.*, 2008).

Natural products have a long history of use in traditional medicine history like Egyptian, Arabic, Chinese and Ayurvedic (Sarker & Nahar, 2007). The natural products have been considered not necessarily reliable to the effective maintenance of good health (Lin & Lin, 1996).

In the present research work, the two methods are used to obtain the extracts from different herbal products. The methods include the Microwave assisted extraction, a modern method of extraction in house hold as well as for scientific research; and the Stove heating extraction, which is a traditional method of extraction. The microwave assisted extraction and the traditional method of extraction i.e. stove heating extraction using the samples including Green Tea, Joshanda, Coffee, Tapal Tez Dam Tea, Tetley Tea, Supreme Tea, Tapal Danedar Tea, Qehwa, Vital Tea and Lipton Tea which were radily available in the market. A comparison was also made between the two methods by studying the extract production in both cases.

In using the samples of Green Tea, Joshanda, Coffee, Tapal Tez Dam Tea, Tetley Tea, Supreme Tea and Tapal Danedar Tea, results showed that Microwave assisted extraction method gave the higher extracts as compared to the Stove heating extraction method. For example, in case of Green Tea the extract produced by the microwave assisted extraction was 1.27g and the extract obtained by stove heating was 1.002g.

Similarly in case of Joshanda, the value of the extract obtained by microwave assisted extraction was 3.76g and by stove heating extraction it was 3.12g. On the other hand the samples including the Qehwa, Vital Tea and Lipton Tea, there was no significant difference between the extracts obtained by microwave assisted extraction and stove heating extraction method. For example, in the extraction of Qehwa the extract obtained by microwave assisted extraction was 1.554g and by stove heating, it was 1.334g.

Microwave Assisted Extraction (MAE) has reported to be an efficient and rapid method for extraction of plant products. In earlier research works when microwave assisted extraction was compared with the conventional extraction methods, it gave high yield, higher extraction selectivity in lesser time. All the other related industries like food, cosmetics and medicinal industries might be promoted from this emerging technology of microwave assisted extraction MAE.

A bio-active phyto-chemical, curcumin, was isolated by using soxhlet extraction, supercritical carbon dioxide assisted extraction and microwave assisted extraction from dried rhizomes of *Curcuma longa and quantity and quality of product was compared*. Microwave assisted extraction was found to be more efficient as well as rapid for the maximum extraction of curcumin using acetone as a solvent (Wakte *et al.*, 2011). In some other experiments, microwave-assisted hydrodistillation wascompared for extraction of essential oils from the aerial parts (tops) of *Thymus vulgaris* L. (common thyme) in terms of extraction yield, extraction time, stability, chemical composition and cost effectiveness. It was found that microwave assisted hydrodistillation was superior as compared to conventional hydrodistillation and was named as green technology (Golmakani & Rezaei, 2008).

In the process of extraction, the benefit of microwave heating is basically the interruption of weak hydrogen bonds (due to dipole rotation). Furthermore due to the production of ions there is production of Ionic currents in electric field. Matrix and solvent resist against these ionic currents, so heat is released by a Joule effect. These all events depend on the dimension and charge of the ions present in that solution.

Solvents can be used individually in microwaves as well as in combinations (Renoe, 1994). Microwave assisted extraction is also more suitable for thermolabile compounds (Romele & Polesello, 1997).

Localized heating caused by microwaves directs towards the expansion and then rupture of cell walls. This causes the more rapid release of secondary metabolites from the cells (Pare´ *et al.*,

1994; Jassie *et al.*, 1997; Majors, 1998). One thing to be considered in microwave assisted extraction is moisture content because water is superheated and then it promotes metabolites to be released in the surrounding medium (Onuska & Terry, 1993; Budzinski *et al.*, 1996; Jassie *et al.*, 1997). In addition, control of the water content of the matrix allows more reproducible results.

Atoui *et al.*, studied different tea and herbal infusions for their polyphenolic content, antioxidant activity and phenolic profile. The total phenolics recovered by ethyl acetate from the water extract, were determined by the Folin–Ciocalteu procedure and ranged from 88.1 ± 0.42 (Greek mountain tea) to 1216 ± 32.0 mg (Chinese green tea) GAE (Gallic acid equivalents)/cup. Anesini *et al.*, analyzed twelve samples of eight brands of tea.

References

Abu-Samara, A., Morris, J. S. and Koirtyohann, S. R. 1975 . Wet ashing of some biological samples in a microwave oven. *Journal of Analytical Chemistry.* 47: 1475-1477.

Alupului, A., Ioan, C. and Lavic, V. 2012 . Microwave assisted extraction of active principles from medicinal plants. Science Bulletin. 74(2): 130-141.

Al-Rasbi. and Khan, S. A. 2013 . Isolation and quantitative estimation of caffeine content in different brands of coffee and tea leaves. *Scholars academic journal of Biosciences.* 1(3): 67-68.

Alan, M. and Iris, M. 2004 . The empire of tea. The overlook press. 32.

Attard, M. T., Watterson, B., Budarin, L. V., Clark, H. J. and Hunt, J. A. (2014). Microwave assisted extraction as an important technology for valorizing Orange waste. *Journal of Chemistry.* 38: 2278-2283.

Barnabas, I. J., Dean, J. R., Fowlis, I. A. and Owen, S. P. 1995 . Extraction of polycyclic aromatic hydrocarbons from highly contaminated soils using microwave energy. Analyst. 120: 1897-1904.

Bayramoglu, B., Serpil, S. and Sumnu, G. 2008 . Solvent free extraction of essential oils from Oregano. *Journal of food engineering.* 88(4): 535-540.

Budzinski, H., Baumard, P., Papineau, A., Wise, S. and Garrigues, P. 1996 . Focused microwave assisted extraction of polycyclic aromatic compounds from standard reference materials, sediments and biological tissues. *Journal of Polycyclic Aromatic Compounds.* 9: 225-232.

Camel, V. and Bermond, A. 1999 . Microwave solvent extraction of environmental samples. Journal of analytical experimental chemistry. 92: 117-135.

Camel, V. 2000. Microwave assisted solvent extraction of environmental samples. *Journal of trends of analytical chemistry.* 19: 229-248.

Camel, V. 2000. Microwave assisted solvent extraction of environmental samples. *Journal of trends in analytical chemistry.* 19: 4.

Chen, L., Jin, H., Ding, L., Li, J., Qu, C. and Zhang, H. 2008. Dynamic Microwave assisted extraction of flavonoids from *Herba epimedii. Journal of separation and purification technology.* 59: 50-57.

Chen, S. S. and Spiro, M. 1994. Study of microwave extraction of essential oil constituents from plant materials. *Journal of microwave electromagnetic energy.* 29: 231-241.

Demesmay, C. and Olle, M. 1993. Utilization des micro-ondes dans les laboratories d'analyse. *Spectra analyse.* 175: 27-32.

Dai, J., Yaylayan, A. V., Raghavan, V. S. G., Pare, J. J. R., Liu, Z. and Belanger, R. M. J. 2001. Influence of operating parameters on the use of the microwave assisted processing for extraction of Azadirachtin-related limonoids from neem under atmospheric pressure conditions. *Journal of agricultural food and chemistry.* 49(10): 4584-4588.

Evans, M. 1994. A guide to herbal remedies. Orient Paperbacks.

Gupta, L. M. and Raina, R. 1998. Side effects of some medicinal plants. *Journal of Current Science.* 109(4): 925-930.

Ganzler, K., Szinai, I. and Salgo, A. 1990. Effective sample preparation for extracting biological active compounds from different mixtures by a microwave technique. *Journal of Chromatography.* 520: 257-262.

Golmakani, T. M. and Rezaei, K. 2008. Comparison of microwave assisted extraction hydrodistillation with traditional hydrodistillation method in the extraction of essential oil from Thymus vulgarisL. *Journal of food chemistry.* 75: 897-900.

Guihua, W., Ping, S., Fan, Z., Yan, X. H., Yi, Y., Zhenku, G. 2011. Comparison of microwave assisted extraction of aloe-emodin in aloe with Soxhlet extraction and ultrasonic assisted extraction. Jornal of China Science Chemistry. 54(1): 231-236.

Handa, S. S., Khanuja, S. P., Longo, G. and Rakesh, D. D. 2008. Extraction technologies for medicinal and aromatic plants. *Journal of International centre for science and high technology.* 1-10.

Jassie, L., Revesz, R., Kierstead, T., Hasty, E. and Matz, S. 1997. Microwave assisted solvent extraction in microwave enhanced chemistry: fundamentals, sample preparation and applications. *Journal of American chemistry society.* 569-609.

Kaptchuk T. J. and Tilburt J. C. 2008. Ethical analysis of Herbal Medicines and Global Health. Bulletin of the World Health Organization. 86: 594-599.

Kamboj, V. P. 2000. Herbal Medicine. *Journal of Current Science.* 78: 35-9.

Lin, G. D. and Lin, G. P. 1996. Chinese medicine, Taipei, Publishers, Taiwan. 188-189.

Lianfu, Z. and Zelang, L. 2008. Optimization and comparision of Ultrasonic/microwave assisted extraction(UMAE) and Utrasonic assisted extraction(UAE) of lycopene from tomatoes. *Journal of ultrasonic sono chemistry.* 15(5): 731-737.

Lucchesi, M. E., Chemat, F. and Smajda, J. 2004. Solvent free microwave extraction of essential oils from aromatic herbs: comparison with conventional hydrodistillation. *Journal of chromatography A.* 1043: 323-327.

Majors, R. E. 1998. Sample preparation perspectives. New approaches to sample preparation. LC-GC Int 8: 128-133.

Onuska, F. I. and Terry, K. A. 1995. Microwave extraction in analytical chemistry of pollutants: polychlorinated biphenyles. *Journal of high resolution chromatography.* 18: 417-421.

Onuska, F. I. and Terry, K. A. 1993. Extraction of pesticides from sediments using a microwave technique. *Journal of Chromatography.* 36: 191-194.

Okoh, O. O., Sadimanko, P. A. and Afolayan, A. J. 2010. Comparative evaluation of the antibacterial activities of the essential oils of *Rosmarinus officinalisL.* Obtained by hydrodistillation and solvent free microwave assisted extraction methods. *Journal of food chemistry.* 120(1): 308-312.

Orio, L., Cravotto, G., Binello, A., Pignata, G., Nicola, S. and Chemat, F. 2012. Hydrodistillation and in-situ microwave generated hydrodistillation of fresh and dried mint leaves: a comparison study. *Journal of science of food and agriculture.* 92(15): 3085-3090.

Pare', J. R. J. 1990. Microwave extraction of volatile oils and apparatus therefore patent no. 90250286. 3 (0 485 668 A1).

Pare', J. R. J. 1991. Microwave assisted natural products extraction. Patent no. 519, 588 (5, 002, 784).

Pare', J. R. J. 1994. Microwave extraction of volatile oils. Patent no. 29, 358 (5, 338, 577).

Pan, X., Niu, G. and Liu, H. 2003. Microwave assisted extraction of tea polyphenoles and tea caffeine from green tea leaves. *Journal of chemical engineering and processing.* 42: 129-133.

Pan, X., Niu, G. and Liu, H. 2002. Comparison of microwave assisted extraction and conventional extraction technique for the extraction of tanshinones from *Saliva miltorrhiza bunge. Jornal of Biochemical Engineering.* 12: 71-77.

Pan, X., Niu, G. and Liu, H. 2001. Microwave assisted extraction of Tashinones from *Salvia miltiorrhiza bunge* with analysis by high performance liquid chromatograpy. *Journal of chromatography A.* 922(1-2): 371-375.

Renoe, B. W. 1994. Microwave assisted extraction. *Journal of analysis of laboratory.* 26: 34-40.

Romele, L. and Polesello, S. 1997. How to minimize the use of solvents. *Journal of Labratorio 2000.* 11: 102-110.

Sarker, S. D. and Nahar, L. 2007. Chemistry for Pharmacy students General, Organic and Natural product Chemistry. England: John Wiley and Sons. 283-359.

Sinquin, A., Gorner, T. and Dellacheie, E. 1993. Lutilisation des micro-ondes enchinie analytique analusis. 21: 1-10.

Soomro, T. A., Mohammad, M. and Aqeel, Z. 2011. Acetyl salicylic acid (aspirin) and antioxidative agents in Joshanda- a herbal (medicine) tea. *Journal of electrochemical society.* 33(26): 1-8.

Thue'ry, J. 1992. Microwaves and matter. In microwaves: industrial, scientific and medical applications, Grant EH(ed). Artech House: London. 1: 83-125.

Wakte, S. P., Sachin, S. B., Patil, A. A., Mohato, M. D., Band, H. T. and Shinde, B. B. 2011. Optimization of microwave, ultrasonic and supercritical carbon dioxide assisted extraction techniques for Curcumin from *Curcuma longa. Journal of separation and purification technology.* 79(1): 50-55.

Winslow, L. C. and Kroll, D. J. 1998. Herbs as medicine. Archieves of Internal Medicines. 158: 2192- 9.